Bond
No.1 for exam success

D0526467

SATs Skills

Arithmetic Workbook

9–10 years

OXFORD
UNIVERSITY PRESS

OXFORD
UNIVERSITY PRESS

Great Clarendon Street, Oxford, OX2 6DP, United Kingdom

Oxford University Press is a department of the University of Oxford.
It furthers the University's objective of excellence in research, scholarship,
and education by publishing worldwide. Oxford is a registered trade mark
of Oxford University Press in the UK and in certain other countries

British Library Cataloguing in Publication Data
Data available

978-0-19-274564-4

10 9 8 7 6 5 4

Paper used in the production of this book is a natural, recyclable product
made from wood grown in sustainable forests. The manufacturing process
conforms to the environmental regulations of the country of origin.

Printed in China

Acknowledgements

Cover illustrations: Lo Cole

Although we have made every effort to trace and contact all copyright
holders before publication this has not be possible in all cases. If notified
the publisher will rectify any error or omissions at the earliest opportunity.

Number

> **Helpful Hint**
>
> Each **digit** in a number has a **place value**. The **place value** tells us the value of that **digit** by its position in the number.
>
> **Example:** The **digit 5** in the number 1 356 892 has a value of **5** ten thousands.
>
> | Millions (M) | Hundred Thousands (HTh) | Ten Thousands (TTh) | Thousands (Th) | Hundreds (H) | Tens (T) | Units (U) |
> |---|---|---|---|---|---|---|
> | 1 | 3 | **5** | 6 | 8 | 9 | 2 |

(A) Answer these questions. You do not need to show your workings out.

1 Which number is bigger, 234 653 or 234 563? _234 653_ ✓ ☐ 1 [1]

2 What is the smallest number you can make using all of these digits?

7 4 8 3 7 6 _346 778_ ✓ ☐ 1 [1]

3 What is the value of the digit 3 in the number 437 785? _30 000_ ✓ ☐ 1 [1]

4 Which number is smaller, 898 989 or 899 898? _898 989_ ✓ ☐ 1 [1]

5 Which digit represents the millions in 5 337 289? _5_ ✓ ☐ 1 [1]

6 Count forward 10^2 from 786. _886_ ✓ ☐ 1 [1]

7 Which number comes next if counting backwards in steps of 10^5?

563 676   463 676 _363 676_ ✓ ☐ 1 [1]

8 Count backwards 10^3 from 786 654. _787 654_ X ☐ 0 [1]

9 Which digit represents the ten thousands in 265 908? _6_ ✓ ☐ 1 [1]

> **Helpful Hint**
>
> 10 to the power of 2 = 10^2 (10 × 10) = 100
>
> 10 to the power of 3 = 10^3 (10 × 10 × 10) = 1000
>
> **4 is a square number.** You can arrange 4 bricks in a perfect square of 2 bricks in length - so 2 × 2
>
> **8 is a cube number.** It can be made by placing 8 bricks in this way, forming a cube with each side being 2 bricks in length.

8 / 9

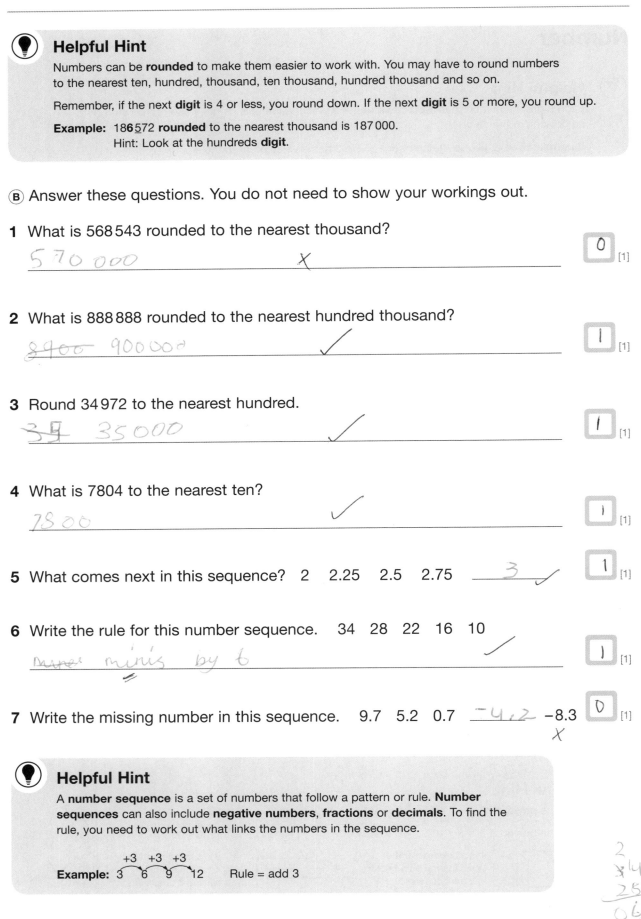

Helpful Hint

Numbers can be **rounded** to make them easier to work with. You may have to round numbers to the nearest ten, hundred, thousand, ten thousand, hundred thousand and so on.

Remember, if the next **digit** is 4 or less, you round down. If the next **digit** is 5 or more, you round up.

Example: 186<u>5</u>72 **rounded** to the nearest thousand is 187 000.
Hint: Look at the hundreds **digit**.

B Answer these questions. You do not need to show your workings out.

1 What is 568 543 rounded to the nearest thousand?

570 000   ✗   0 [1]

2 What is 888 888 rounded to the nearest hundred thousand?

~~8400~~ 900 000   ✓   1 [1]

3 Round 34 972 to the nearest hundred.

~~34~~ 35 000   ✓   1 [1]

4 What is 7804 to the nearest ten?

7800   ✓   1 [1]

5 What comes next in this sequence? 2 2.25 2.5 2.75 __3__ ✓   1 [1]

6 Write the rule for this number sequence. 34 28 22 16 10

~~numer~~ minis by 6   ✓   1 [1]

7 Write the missing number in this sequence. 9.7 5.2 0.7 __-4.2__ –8.3   0 [1]
✗

Helpful Hint

A **number sequence** is a set of numbers that follow a pattern or rule. **Number sequences** can also include **negative numbers**, **fractions** or **decimals**. To find the rule, you need to work out what links the numbers in the sequence.

Example: 3 6 9 12   Rule = add 3
(+3 +3 +3)

2
×4
25
06

5 / 7

Ⓒ Answer these questions. You do not need to show your workings out.

1 Write the factors of 12.

1, 12, 6, 2, 4, 3   ✓   [1] 1

2 What are the common factors of 12 and 18?

1, ~~4~~ 3, 6   ✗   [1] 0

3 Write the factor pairs of 36.

(1, 36)(9, 4)(18, 2)(6 × 6)   ✗   [1] 0

4 Write a prime number between 6 and 10.

~~5~~ 7   ✓   [1] 1

5 Write the prime factors of 21.

7, 3   ✓   [1] 1

6 How many different prime factors does 84 have?

3   ✗   [1] 0

3 / 6

Word problems

(D) Solve these word problems and show your workings out.

1 A huge outdoor music concert has been organised. 33 546 tickets have been sold. The number of tickets sold was rounded to the nearest 100 and to the nearest 10.

Which number is bigger?

33 546 = (10) 33
= 33 550
(100) = 33 500

33 550 ✓
to

[1]

2 It is cold outside! The thermometer reads –1°C. The temperature is dropping 2°C each hour.

What will the temperature be in 2 hours' time?

$-1°C - 2°C - 2°C = -5°C$

$-5°$ ✓

[1]

3 Add together all the prime numbers between 6 and 20.

Does the answer make a prime number?

7 +
$7 + 11 + 13 + 17 + 19 = 67$

67 ✓
yes

[1]

4 Tuhil threw a dice six times. He got the numbers 6, 4, 1, 3, 4, 5.

What are the smallest and the largest numbers he can make from these digits?

65, 13

a 13   **b** 65

✗

0 [2]

5 Ben has 12 buttons and places them in a line to represent the factor pair (1, 12) 1 × 12. What other factor pairs of 12 can he make if he places the buttons in rectangle shapes?

2,6, 4 3

(2,6)(4,3)

✗

0 [1]

3 / 6

Addition

(💡) **Helpful Hint**

To work out the answers as quickly as possible, it is sometimes helpful to add the thousands first, followed by the hundreds, tens and units.

(A) Answer these questions. You do not need to show your workings out.

1 5000 + 5064 = _10064_ ✓ |1| [1]

2 23003 + 6003 = _29006_ ✓ |1| [1]

3 4890 + 10110 = _~~44~~ 15000_ ✓ |1| [1]

4 6666 + 4444 = _11 110_ ✓ |1| [1]

5 54000 + 46000 = _114 000_ ✗ |0| [1]

6 250 + 2500 + 2750 = _5500_ ✓ |1| [1]

7 4498 + 3502 + 2000 = _10 000_ ✓ |1| [1]

8 16563 + 2500 = _19 663_ ✗ |0| [1]

9 25497 + 27670 = _54 567_ ✗ |0| [1]

10 4624 + 192 + 1983 = _6799_ ✓ |1| [1]

7 / 10

Helpful Hint

Remember, the numbers in an addition calculation can be added together in any order but you must keep the units, tens, hundreds and thousands under their correct column headings.

Example:

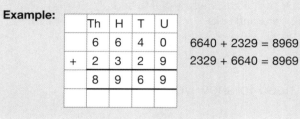

Th	H	T	U
	6	6	4	0
+	2	3	2	9
	8	9	6	9

6640 + 2329 = 8969

2329 + 6640 = 8969

(B) **Answer these questions and show your workings out.**

1 7734 + 6529 =

O [1]

2 12838 + 6634 =

19472 ✓ [1]

3 34997 + 26265 =

6126 ✓ [1]

4 66363 + 87212 =

153575 ✓ [1]

5 35034 + 6709 =

41743 ✓ [1]

4 / 5

6 73 488 + 21 221 =

```
    7 3   4 8 8
    2 1   2 2 1
    9 4   7 0 9
```

94 709 ✓   1 [1]

7 43 328 + 23 899 =

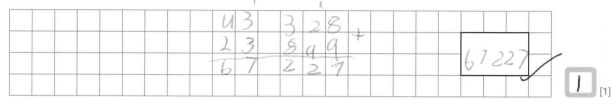

```
    4 3   3 2 8 +
    2 3   8 9 9
    6 7   2 2 7
```

67 227 ✓   1 [1]

8 78 236 + 38 993 + 4 533 =

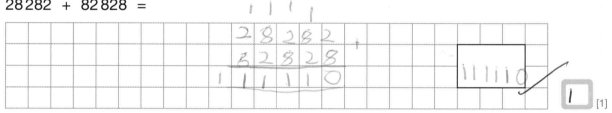

```
    7 8   2 3 6
    3 8   9 9 3
     4   5 3 3
  1 2 7  7 6 2
```

121 762 ✓   1 [1]

9 28 282 + 82 828 =

```
    2 8 2 8 2 +
    8 2 8 2 8
  1 1 1 1 1 0
```

111 110 ✓   1 [1]

10 17 698 + 12 440 + 9 072 =

```
    1 7   6 9 8
    1 2   4 4 0 +
     9   0 7 2
    3 9   2 2 0
```

39 220 ✗   0 [1]

11 99 427 + 709 + 8 560 =

```
    9 9   4 2 7
         7 0 9 +
     8   5 6 0
  1 0 8  6 9 6
```

108 696 ✗   0 [1]

4 / 6

Unit 2

Word problems

 Helpful Hint

Remember **decimal points** are used in money to separate the pounds from the pence.

It is important to use columns when adding numbers with decimals. You need to line up the pounds (£), **decimal points** and pence.

Example:

	£	4	2	.	3	1
+	£		6	.	1	2
	£	4	8	.	4	3

ⓒ Solve these word problems and show your workings out.

1 A supermarket placed a large order for jam. They ordered 4750 jars of strawberry jam, 3200 jars of blackcurrant jam, 4230 jars of raspberry jam and 1665 jars of apricot jam.

How many jars were ordered in total?

4750
3200
4230 +
1665
13 8 45

13845 ✓

1 [1]

2 Oka spent £12.75, £108.95 and £9.99 during a shopping trip.

How much did she spend altogether?

£108.95
£012.75 +
£009.99
£131.69

£12.75
£08

£131.69 ✓

1 [1]

3 A DJ played three songs. The songs lasted 3 minutes and 11 seconds, 2 minutes and 54 seconds, and 3 minutes and 18 seconds.

How long in total did it take to play all three songs? Remember, there are 60 seconds in 1 minute.

3 + 3 + 2 = 8 min
54 + 18 + 11 = 83 sec = 1 min 23 sec

9 min 23 sec ✓

1 [1]

4 Anna had £100 to spend. She booked two horse-riding lessons, costing £35.75 each, and a jumping lesson costing £45.25.

How much more than £100 did the lessons cost?

35.75 + 45.25 = £81

£81

✗

0 [1]

3 / 4

Subtraction

Helpful Hint

Subtracting one number from another is the same as finding the difference between them. You can use a **number line** to help work this out.

Example: $532 - 510 = 22$

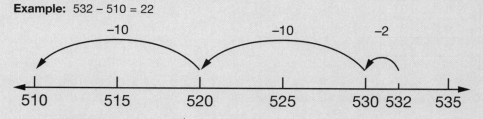

| −10 | −10 | −2 |

| 510 | 515 | 520 | 525 | 530 532 | 535 |

The **number line** shows that the difference between 510 and 532 is 22.

Ⓐ Answer these questions. You do not need to show your workings out.

1 $5600 - 2300 =$ _3300_ ✓ | | 1 [1]

2 $10000 - 8450 =$ _1550_ ✓ | | 1 [1]

3 $554 - 55 =$ ~~609~~ 449 ✓ | | 1 [1]

4 $23754 - 1200 =$ _22 554_ ✓ | | 1 [1]

5 $18995 - 6990 =$ _12 005_ ✓ | | 1 [1]

6 $34287 - 3200 =$ _31 087_ ✓ | | 1 [1]

7 $45700 - 5642 =$ _40 142_ ✗ | | 0 [1]

8 $877 - 98 =$ _779_ ✓ | | 1 [1]

9 $20000 - 648 =$ _19 352_ ✓ | | 1 [1]

10 $37590 - 475 =$ _37 115_ ✓ | | 1 [1]

9 / 10

Helpful Hint

Remember you need to write the larger number above the smaller number when doing a **subtraction** calculation. Always start with the units, then subtract the tens, then the hundreds and so on.

The units, tens, hundreds and thousands need to be lined up in their correct columns.

Example:

Th	H	T	U
6	4	3	2
− 4	4	2	1
2	0	1	1

Sometimes you need to 'borrow' from the next column.

Example:

Th	H	T	U
7	⁴5̷	¹3	9
− 1	2	7	2
6	2	6	7

We cannot subtract 7 from 3 so we 'borrow' 10 tens (1 hundred) from the hundreds column. Now we can subtract 7 from 13.

Ⓑ Answer these questions and show your workings out.

1 8796 − 1767 =

7029 ✓ [1]

2 10000 − 4567 =

5 433 ✓ [1]

3 54832 − 12826 =

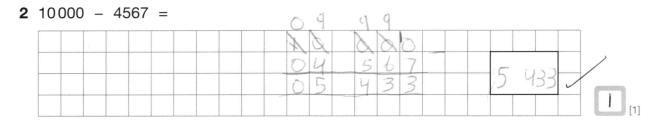

42008 ✓ [1]

4 66000 − 9867 =

56133 ✓ [1]

4 / 4

5 27549 − 977 =

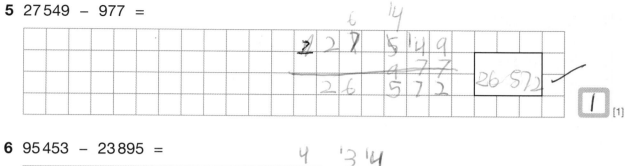

26 572 ✓ | **1** [1]

6 95453 − 23895 =

71 558 ✓ | **1** [1]

7 446633 − 33333 =

413 300 ✓ | **1** [1]

8 120000 − 74553 =

45 447 ✓ | **1** [1]

9 229897 − 129978 =

100 119 ✗ | **0** [1]

10 872123 − 434466 =

437 657 ✓ | **1** [1]

5 / 6

Unit 3

Word problems

Helpful Hint

1 litre is equal to 1000 ml. There are 10 mm in 1 cm and 100 mm in 10 cm. There are 1000 m in 1 km.

ⓒ Solve these word problems and show your workings out.

1 Jay made 3 litres of drink. How much water did he add to 637 ml of blackcurrant squash concentrate?

[1]

2 £10 865 was raised at a school fair. The cost of running the fair was £2453.

How much profit was made?

[1]

3 The Golding family spent £32.65 at a fast food restaurant.

How much change did they get from £50?

[1]

4 A builder cut two lengths of piping. One was 1076 mm long and the other was 155 cm long.

What was the difference in length between the two pieces of piping?

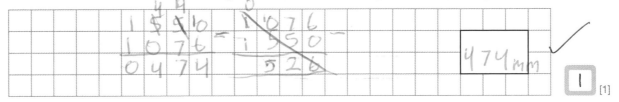

[1]

4 / 4

Multiplying and dividing by 10, 100, 1000

💡 **Helpful Hint**

When you multiply a number by 10, you move all its **digits** one place to the left.

When you multiply a number by 100, you move all its **digits** two places to the left.

When you multiply a number by 1000, you move all its **digits** three places to the left.

We write zeros to make the new places clear as the number becomes bigger.

Whole numbers	**Decimals**
Examples: $563 \times 10 = 5630$	$56.3 \times 10 = 563$
$563 \times 100 = 56\,300$	$56.3 \times 100 = 5630$
$563 \times 1000 = 563\,000$	$56.3 \times 1000 = 56\,300$

When multiplying whole numbers by 10, 100, 1000 and so on you take the number of zeros and place them at the end of the number you are multiplying.

Example: $563 \times 100 = 56300$

Remember, we do not need to add a .0 to a whole number. Whole numbers are followed by an invisible **decimal point**. For example, write 563 not 563.0.

Ⓐ Answer these questions. You do not need to show your workings out.

1 $356 \times 100 = 35600$ ✓   [1] [1]

2 $298 \times 10 = 2980$ ✓   [1] [1]

3 $83 \times 1000 = 83\,000$ ✓   [1] [1]

4 $5632 \times 100 = 563\,200$ ✓   [1] [1]

5 $99.9 \times 1000 = 99\,900$ ✓   [1] [1]

6 $7753 \times 10 = 77\,530$ ✓   [1] [1]

7 $1010 \times 100 = 101\,000$ ✓   [1] [1]

8 $650 \times 1000 = 650\,000$ ✓   [1] [1]

9 $72.3 \times 10 = 723$ ✓   [1] [1]

10 $487 \times 100 = 48\,700$ ✓   [1] [1]

10 / 10

Helpful Hint

When you divide a number by 10, you move all its **digits** one place to the right.

When you divide a number by 100, you move its all **digits** two places to the right.

When you divide a number by 1000, you move its all **digits** three places to the right.

Examples: $563\,000 \div 10 = 56\,300$

$563\,000 \div 100 = 5630$

$563\,000 \div 1000 = 563$

When working with **decimal numbers** the **digits** move in the same way.

$56.3 \div 10 = 5.63$

$56.3 \div 100 = 0.563$

$56.3 \div 1000 = 0.0563$

Remember to put a **decimal point** if it is needed, and a 0 before the decimal point.

(B) Answer these questions and show your workings out.

1 $780 \div 10 =$

78 ✓

1 [1]

2 $5600 \div 100 =$

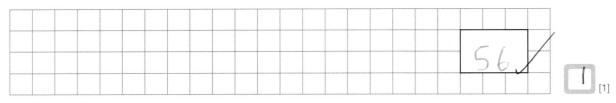

56 ✓

1 [1]

3 $23\,000 \div 1000 =$

23 ✓

1 [1]

4 $126.9 \div 100 =$

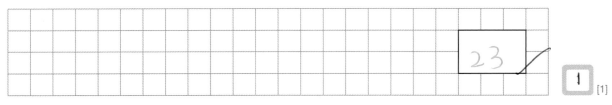

1.269 ✓

1 [1]

Unit 4

ⓒ Answer these questions and show your workings out.

1 943 ÷ 100 =

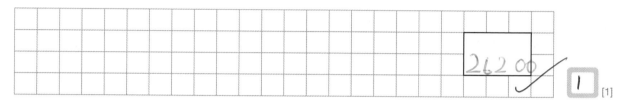

9.43 ✓ | 1 | [1]

2 26.2 × 1000 =

262 00 ✓ | 1 | [1]

3 4 ÷ 10 =

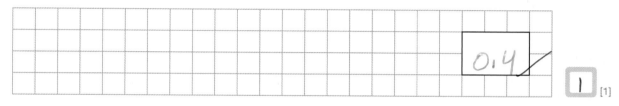

0.4 ✓ | 1 | [1]

4 770 × 100 =

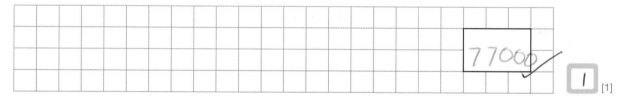

77000 ✓ | 1 | [1]

5 23.67 × 100 =

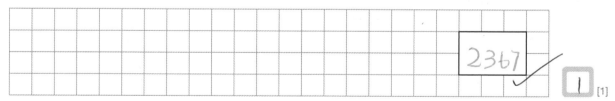

2367 ✓ | 1 | [1]

6 5.72 ÷ 10 =

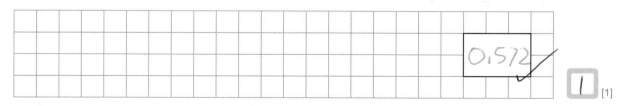

0.572 ✓ | 1 | [1]

6 / 6

Word problems

What is the units?

(D) Solve these word problems and show your workings out.

1 4800 tins of baked beans were packed into 100 boxes.
How many tins are in each box?

$4800 \div 100 = \boxed{48}$ ✓

$\boxed{1}$ [1]

2 A seagull flew 0.37 km above the sea.
How many metres above the water was it flying?

$0.37 = \boxed{370}$ ✓

$\boxed{1}$ [1]

3 Dan's mobile phone package included 300 free texts a month.
Each additional text would cost 10p. Dan sent 456 texts in June.
How much extra did Dan have to pay for his texts?

$456 - 300 \times 10 \boxed{1560}$ ✗

$\boxed{0}$ [1]

4 A plane flew a distance of 13 600 km in 1000 minutes.
If it travelled the same distance every minute, how many kilometres did it travel in one minute?

$13600 \div 1000 = \boxed{13.6}$ ✓

$\boxed{1}$ [1]

5 Frans has fifteen €100 notes. Tickets to a theme park cost ten euros each.
How many tickets could he buy?

$500 \div 10 = \boxed{50 \text{ tickets}}$ ✗

$\boxed{0}$ [1]

$\boxed{3} \Big/ 5$

Multiplication

> **Helpful Hint**
>
> When you square a number, you multiply it by itself.
>
> **Example:** 5×5 is the same as 5^2. Both give the answer 25.
>
> When you cube a number, you multiply it by itself and then by itself again.
>
> **Example:** $5 \times 5 \times 5$ is the same as 5^3. Both give the answer 125.
>
> The **square** and **cube number** of an even number is always even.
>
> The **square** and **cube number** of an odd number is always odd.

A Answer these questions. You do not need to show your workings out.

1 a $3^2 =$ ___9___ ✓ **b** $3^3 =$ ___27___ ✓ **2** [2]

2 a $2^2 =$ ___4___ ✓ **b** $2^3 =$ ___8___ ✓ **2** [2]

3 a $4^2 =$ ___16___ ✓ **b** $4^3 =$ ___64___ ✓ **2** [2]

4 a $10^2 =$ ___100___ ✓ **b** $10^3 =$ ___1000___ ✓ **2** [2]

5 $3^2 + 2^3 =$ ___17___ ✓ **1** [1]

6 $9^2 + 7^3 =$ ___361___ ✗ **0** [1]

7 $6^2 + 10^3 =$ ___1036___ ✓ **1** [1]

8 $8^2 + 1^3 =$ ___65___ ✓ **1** [1]

11 / 12

Helpful Hint

Remember to multiply each number in turn, starting with the units (on the right). If the result is 10 or more, you need to carry the tens **digit** to the next column to the left.

Example: 2387 × 4 = ?

Th	H	T	U
2	3	8	7
×			4
9	5	4	8
1	3	2	

Step 1 7 × 4 = 28, so carry the 2 to the tens column. The 8 stays in the units column.

Step 2 8 × 4 = 32. 32 + the 2 carried over = 34. The 3 is carried over to the hundreds column and the 4 stays in the tens column.

Step 3 3 × 4 = 12. 12 + the 3 carried over = 15. The 1 is carried over to the thousands column and the 5 stays in the hundreds column.

Step 4 2 × 4 = 8. 8 + the 1 carried over = 9.

(B) Answer these questions and show your workings out.

1 4372 × 3 =

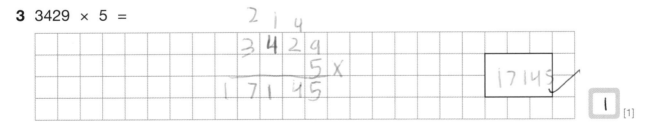

13116 ✓ 　1 [1]

2 7878 × 6 =

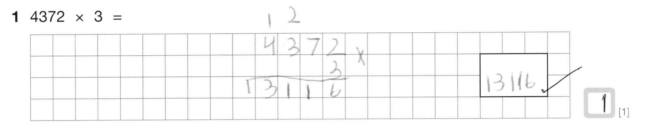

47268 ✓ 　1 [1]

3 3429 × 5 =

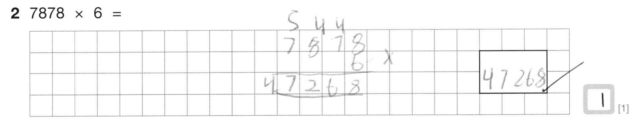

17145 ✓ 　1 [1]

4 6788 × 8 =

57 504 ✗ 　0 [1]

3 / 4

Helpful Hint

Long multiplication is a way of writing a multiplication when the **multiplier** has more than one **digit**.

Example: 253 × 27 = ?

		2	5	3
	×		2	7
1	7₃	7₂	1
5₁	0	6	0
6	8₁	3	1

Step 1 Multiply each **digit** in the top number by 7. This will give us 1771.

Step 2 Multiply the top number by 20. 20 = 2 × 10. Write a 0 in the units column so the **digits** are moved one place to the left (multiplied by 10). Then multiply by 2. 3 × 2 = 6, so write the 6 in the correct (tens) column.

Step 3 Then multiply 5 × 2 = 10. The 1 gets carried to the thousands column and the 0 goes in the hundreds column.

Step 4 Next multiply 2 × 2 = 4. Add on the 1 we carried to make 5, this goes in the thousands column. This gives us 5060.

Step 5 Finally add the two numbers together using column addition. 1771 + 5060 = 6831

(c) Answer these questions and show your workings out.

1 346 × 23 =

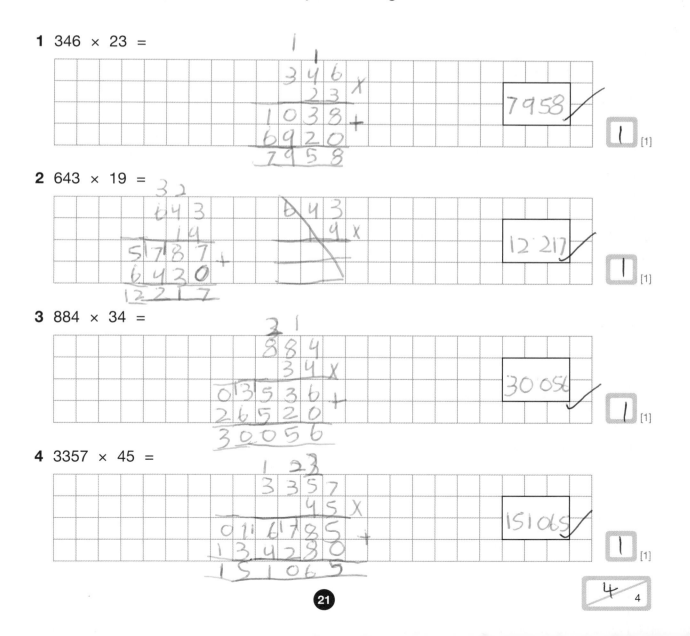

2 643 × 19 =

3 884 × 34 =

4 3357 × 45 =

Unit 5

Word problems

(D) Solve the word problems and show your workings out.

1 If half of a number is 5649, what number did we start off with?

$$5\ 649$$
$$\underline{2\ \times}$$
$$11\ 298$$

$\boxed{11\ 298}$ ✓ $\boxed{l}$ [1]

2 Joe gave half of his marbles to Ali. Then Joe lost 20 more. He now has 656 marbles.

How many did he start with?

$$676$$
$$\underline{2\ \times}$$
$$1\ 352$$

$\boxed{1\ 352}$ ✓ $\boxed{l}$ [1]

3 A farmer had 625 sheep that gave birth to lambs. 415 sheep had twins, 110 had triplets and the rest had single lambs.

How many lambs were born in total?

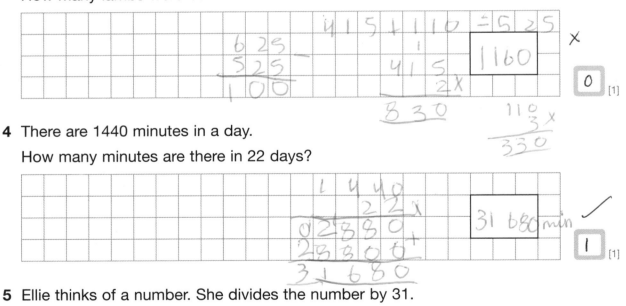

$$625$$
$$\underline{525\ -}$$
$$100$$

$$415 + 110 = 525$$
$$\boxed{1\ 160}\ \times$$

$$415$$
$$\underline{2\ \times}$$
$$830$$

$$110$$
$$\underline{3\ \times}$$
$$330$$

$\boxed{0}$ [1]

4 There are 1440 minutes in a day.

How many minutes are there in 22 days?

$$1\ 440$$
$$\underline{22\ \times}$$
$$02\ 880$$
$$28\ 300\ +$$
$$31\ 680$$

$\boxed{31\ 680\ min}$ ✓ $\boxed{l}$ [1]

5 Ellie thinks of a number. She divides the number by 31.

The answer is 1250.

What is the number she first thought of?

$$1\ 250$$
$$\underline{31\ \times}$$
$$01\ 250\ +$$
$$37\ 500$$
$$38\ 750$$

$\boxed{38\ 750}$ ✓ $\boxed{l}$ [1]

$\boxed{4}$ / 5

29/08
45m
15m

Division

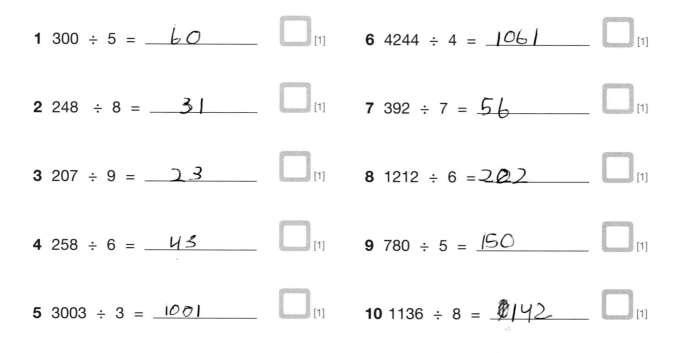

> ### Helpful Hint
>
> Division splits, or shares, amounts into equal groups.
>
> If you are dividing by 4, you can halve then halve again. Or if you are dividing by 8, you can halve, halve and halve again.
>
> **Example:** What is $16 \div 4$?
>
> Half of 16 is 8.
>
> Half of 8 is 4.
>
> So $16 \div 4 = 4$.
>
> Remember that you can also use the 'bus stop' method to work out your answer.
>
> **Example:** $175 \div 7 = ?$
>
> Place the 175 in the 'bus stop'.
>
> $7\overline{)175}$
>
> Look at the first two **digits** (17) and divide them by the 7.
> 7 goes into 17 two times ($2 \times 7 = 14$) with a **remainder** of 3 (this means 3 is left over).
> So the 3 is carried to the next column to make it 35.
>
> $7\overline{)17^{3}5}$ (2)
>
> Next divide 35 by 7 ($5 \times 7 = 35$).
>
> $7\overline{)17^{3}5}$ (2 5)
>
> This gives us the answer, $175 \div 7 = 25$.

Ⓐ Answer these questions. You do not need to show your workings out.

1 $300 \div 5 = $ _60_ ☐ [1]

2 $248 \div 8 = $ _31_ ☐ [1]

3 $207 \div 9 = $ _23_ ☐ [1]

4 $258 \div 6 = $ _43_ ☐ [1]

5 $3003 \div 3 = $ _1001_ ☐ [1]

6 $4244 \div 4 = $ _1061_ ☐ [1]

7 $392 \div 7 = $ _56_ ☐ [1]

8 $1212 \div 6 = $ _202_ ☐ [1]

9 $780 \div 5 = $ _150_ ☐ [1]

10 $1136 \div 8 = $ _142_ ☐ [1]

10

29/08
10m

> **Helpful Hint**
>
> Sometimes a number can't be divided exactly and there is an amount left over in the answer. This is known as the **remainder** (r).
>
> **Example:** $170 \div 6 = 28 \text{ r } 2$
>
> $$6\overline{)1\ ^{1}7\ ^{5}0}$$ (with 2 8 r 2 above) is the same as 170 shared equally between 6 groups = 28 in each group with 2 left over.
>
> Always make sure that you include the **remainders** in your answer.

B Answer these questions and show your workings out.

1 $618 \div 5 =$

$$5\overline{)6\ ^{1}1\ 8}$$ with 1 2 3 r 3 above

123 r 3

[1]

2 $597 \div 7 =$

$$7\overline{)5\ ^{5}9\ ^{3}7}$$ with 0 8 5 r 2 above

85 r 2

[1]

3 $530 \div 4 =$

$$4\overline{)5\ ^{1}3\ ^{1}0}$$ with 1 3 2 r 2 above

132 r 2

[1]

4 $1096 \div 3 =$

$$3\overline{)1\ ^{1}0\ ^{1}9\ 6}$$ with 0 3 6 5 r 1 above

365 r 1

[1]

5 $392 \div 6 =$

$$6\overline{)3\ ^{3}9\ ^{3}2}$$ with 0 6 5 r 2 above

65 r 2

[1]

5

> ## 💡 Helpful Hint
>
> **Remainders** can be written as **fractions**.
>
> The **remainder** is the top number (**numerator**) of the **fraction**. The number you are dividing by is the bottom number (**denominator**).
>
> **Example:** $506 \div 9 = 56 \text{ r } 2$ or $56\frac{2}{9}$
>
> Try to answer the questions on this page using **fractions** for the **remainders**.

6 4196 ÷ 8 =

```
    0 5 2 4 r 4
8 [ 4 4¹ 9³ 6
```

524r4

[1]

7 1427 ÷ 6 =

```
    0 2 3 7 r 5
6 [ 1 ¹4 2² 4⁷
```

237r5

[1]

8 3797 ÷ 9 =

```
    0 4 2 1 r 8
9 [ 3 ³7 4 ¹7
```

421r8

[1]

9 2962 ÷ 7 =

```
    0 4 2 3 r 1
7 [ 2 ²9 ⁶2
```

423r1

[1]

10 571 ÷ 4 =

```
    1 4 2 r 3
4 [ 5 ¹7 ¹1
```

142r3

[1]

5

29/08
15m

Word problems

ⓒ Solve these word problems and show your workings out.

1 Zofia's mum works five days a week and travels a total of 165 miles between home and work each week.

How far does she travel on each working day?

3 2
1 6 5 ×
5
8 2 5

825

[1]

2 Seven children find 204 chocolate eggs in a treasure hunt. The eggs are shared equally between the children.

a How many eggs did each child get?

b How many eggs were left over?

0 2 9
7) 2 0⁶ 4

a 29   b 1

[2]

3 Ice lollies were handed out to all the children at Parklands Community School on sports day. 1260 children received an ice lolly. There were eight ice lollies in each box.

a How many boxes of ice lollies were needed?

b How many ice lollies were left over?

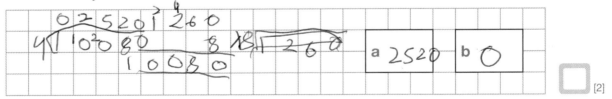

a 2520   b 0

[2]

4 1668 g of cheese was cut into equal pieces and put into six packets.

How much did each packet of cheese weigh?

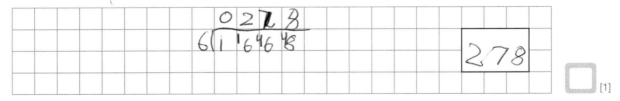

0 2 7 8
6) 1 ⁱ6 ⁴6 ⁸8

278

[1]

Multiplication and division

29/08
15m

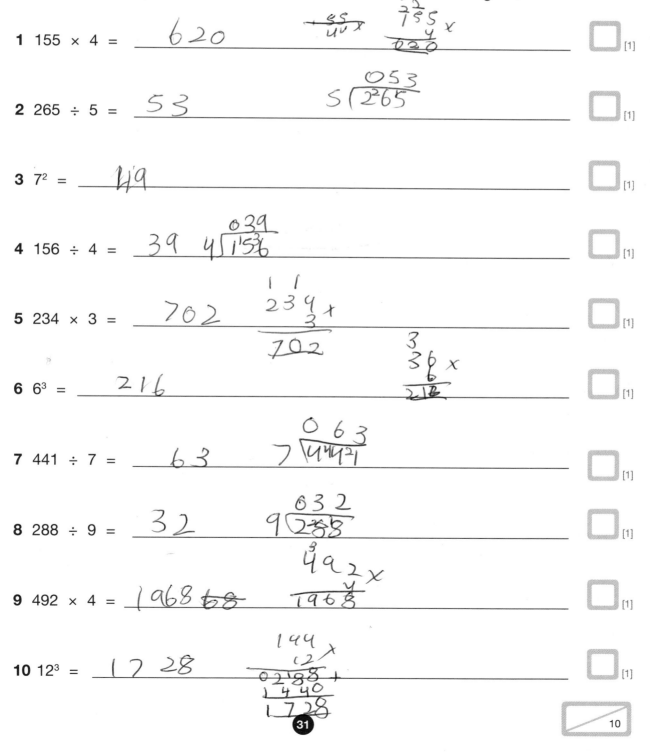

Helpful Hint

To check an answer you can always try its **inverse operation**. The **inverse operation** for multiplication is division and the **inverse operation** for division is multiplication.

Example: $520 \div 8 = 65$

$65 \times 8 = 520$

Try checking your answers by doing the **inverse operation** for some of the questions on this page.

(A) Answer these questions. You do not need to show your workings out.

1 $155 \times 4 =$ ___620___ [1]

2 $265 \div 5 =$ ___53___ [1]

3 $7^2 =$ ___49___ [1]

4 $156 \div 4 =$ ___39___ [1]

5 $234 \times 3 =$ ___702___ [1]

6 $6^3 =$ ___216___ [1]

7 $441 \div 7 =$ ___63___ [1]

8 $288 \div 9 =$ ___32___ [1]

9 $492 \times 4 =$ ___1968___ [1]

10 $12^3 =$ ___1728___ [1]

10

Ⓑ Answer these questions and show your workings out. For the division questions, write the answers with remainders as fractions.

1 7723 × 4 =

7723
4 ×
30892

36892 [1]

2 1706 ÷ 3 =

3 | 1706   0568 r⅔

568 r⅔ [1]

3 2635 × 18 =

2635
18 ×
21080 +
26350
47430

47 430 [1]

4 4463 ÷ 5 =

5 | 4463   0892 r⅗

892 r⅗ [1]

5 3776 × 8 =

3776
8 ×
17268

17 268 [1]

6 5409 ÷ 9 =

9 | 5409   0601

661 [1]

31/08
6 min.

7 3027 ÷ 7 =

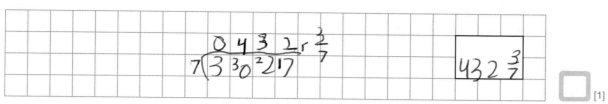

$432\frac{3}{7}$

[1]

8 751 × 23 =

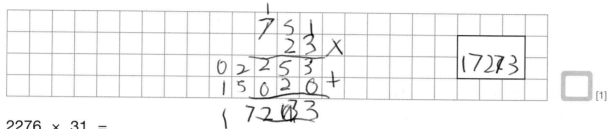

17273

[1]

9 2276 × 31 =

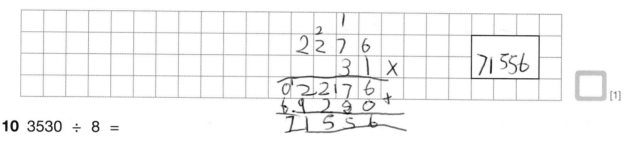

71556

[1]

10 3530 ÷ 8 =

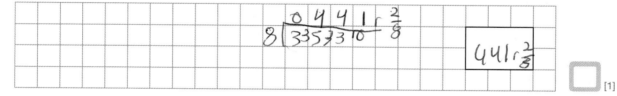

$441 r \frac{2}{8}$

[1]

11 9889 × 44 =

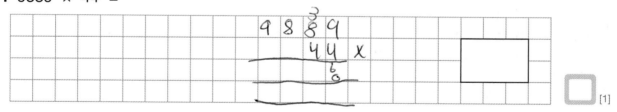

[1]

12 8720 ÷ 4 =

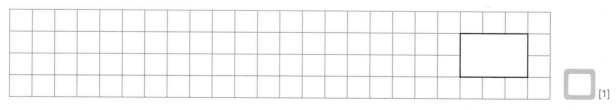

[1]

6

Word problems

ⓒ Solve these word problems and show your workings out.

1 Sundeep's flute lessons cost £17 for half an hour.
How much did 22 hours of lessons cost?

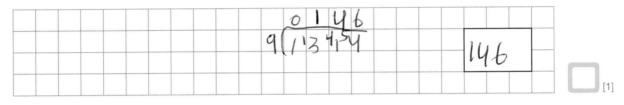

374

[1]

2 A pair of numbers multiplied together makes 1314.
If one of the numbers is 9, what is the other?

146

[1]

3 If a bucket holds 3.5 litres and a cup holds 70 ml, how many cups would be needed to fill the bucket?

$3.5 \div 70ml = 50 cups$ 50

[1]

4 Kingfisher Junior School pupils are going on a trip to Conway.
The trip costs £34.50 for each child. 17 children are going.
How much did the trip cost in total?

$34.50 \times 17 = £586.50$

£586.50

[1]

5 Rani buys 17 DVDs which cost £8.49 each.
How much is this altogether? Give your answer to the nearest £10.

$8.49 \times 17 = £148.49$

6140

[1]

5

31/08
5 min.

Fractions

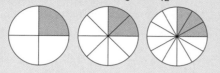

💡 **Helpful Hint**

Remember, **fractions** are used to show parts of a whole. **Equivalent fractions** are written differently but they have the same value.

Example: $\frac{1}{4}$ is exactly the same as $\frac{2}{8}$ and $\frac{3}{12}$.

To find an **equivalent fraction,** you can multiply the **numerator** and **denominator** by the same amount. Let's take the example of $\frac{1}{4}$ used above.

• **Multiply** the numerator and the **denominator** by 2. $1 \times 2 = $ **2** and $4 \times 2 = $ **8**. So the first **equivalent fraction** to $\frac{1}{4}$ is $\frac{2}{8}$.

• Next multiply the **numerator** and **denominator** by 3. $1 \times 3 = $ **3** and $4 \times 3 = $ **12**. So the second **equivalent fraction** to $\frac{1}{4}$ is $\frac{3}{12}$.

• To find the next one, we of course multiply the **numerator** and **denominator** by 4. $1 \times 4 = $ **4** and $4 \times 4 = $ **16**. This means the next **equivalent fraction** of $\frac{1}{4}$ is $\frac{4}{16}$.

Ⓐ Find **one** equivalent fraction for each fraction.

1 $\frac{1}{2}$ = $\frac{4}{8}$ [1]

2 $\frac{2}{3}$ = $\frac{4}{6}$ [1]

3 $\frac{1}{8}$ = $\frac{2}{16}$ [1]

4 $\frac{2}{5}$ = $\frac{4}{10}$ [1]

Find the **first two** equivalent fractions for these fractions.

5 $\frac{1}{7}$ = $\frac{2}{14}$ $\frac{3}{21}$ [2]

6 $\frac{3}{4}$ = $\frac{6}{8}$ $\frac{12}{16}$ [2]

7 $\frac{4}{5}$ = $\frac{26}{25}$ $\frac{8}{10}$ [2]

8 $\frac{1}{9}$ = $\frac{2}{18}$ $\frac{3}{27}$ [2]

12

31/08
3 min

💡 Helpful Hint

Sometimes a **fraction** might have to be simplified to find the smallest possible **numerator** and **denominator**.

Example: What is $\frac{18}{42}$ in its **simplest form**?

First, we need to find the **common factors** of 18 and 42. A **factor** is a whole number which divides into another whole number exactly. We need to find **factors** which divide into both 18 and 24.

1, 2, 3 and 6 can all be divided into 18 and 42. The highest **common factor** here is 6.

18 ÷ 6 = **3**   42 ÷ 6 = **7**

The answer is $\frac{3}{7}$.

B Simplify these fractions and write the answers in their simplest form. Show your workings out.

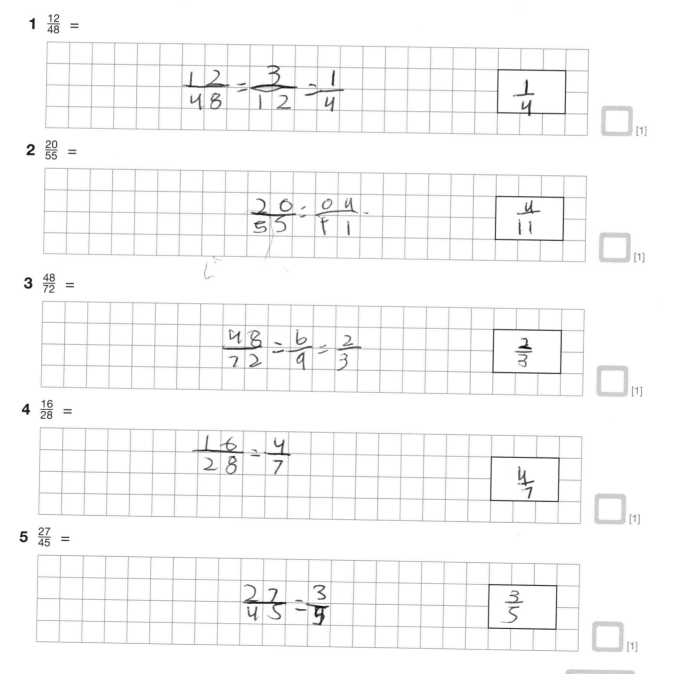

1 $\frac{12}{48}$ =

$\frac{12}{48} = \frac{3}{12} = \frac{1}{4}$ $\boxed{\frac{1}{4}}$ [1]

2 $\frac{20}{55}$ =

$\frac{20}{55} = \frac{04}{11}$ $\boxed{\frac{4}{11}}$ [1]

3 $\frac{48}{72}$ =

$\frac{48}{72} = \frac{6}{9} = \frac{2}{3}$ $\boxed{\frac{2}{3}}$ [1]

4 $\frac{16}{28}$ =

$\frac{16}{28} = \frac{4}{7}$ $\boxed{\frac{4}{7}}$ [1]

5 $\frac{27}{45}$ =

$\frac{27}{45} = \frac{3}{5}$ $\boxed{\frac{3}{5}}$ [1]

 5

3/08
3 min

Helpful Hint

An **improper fraction** means the **numerator** is larger than the **denominator**.

Example: $\frac{12}{6}$

A **mixed number** is a mixture of a **whole number** and a **fraction**.

Example: $3\frac{2}{5}$

To change an **improper fraction** to a **mixed number,** you need to divide the **numerator** by the **denominator**. This tells you the **whole number** part. The **remainder** stays as a **fraction**.

Example: Write $\frac{28}{3}$ as a **mixed number**.

First we divide the **numerator** by the **denominator**: $28 \div 3 = 9 \text{ r } 1$

The 9 is the whole number and the r 1 means $\frac{1}{3}$ is left over.
So the answer is:

Numerator

Whole number $\longrightarrow 9\frac{1}{3}$

Denominator

---

c Write these improper fractions as mixed numbers. Show your workings out.

1 $\frac{16}{5}$ =

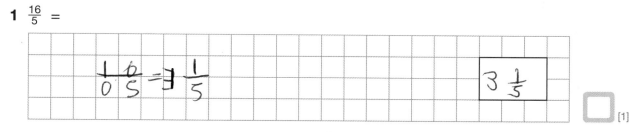

$3\frac{1}{5}$ [1]

2 $\frac{56}{9}$ =

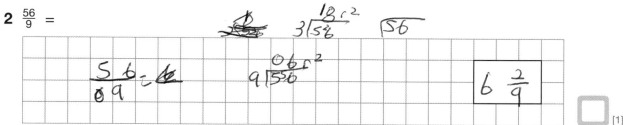

$6\frac{2}{9}$ [1]

3 $\frac{23}{6}$ =

$3\frac{5}{6}$ [1]

4 $\frac{37}{4}$ =

$9\frac{1}{4}$ [1]

4

Word problems

> 💡 **Helpful Hint**
>
> Remember, when adding **fractions** together if the **denominators** are the same, you just need to add the **numerators** together and the **denominator** remains the same.

ⓓ Solve these word problems and show your workings out.

1 Toby, Rowan and Aleksy each bought the same pizza. Toby said he ate $\frac{1}{3}$ of his pizza, Rowan said he ate $\frac{3}{9}$ of his pizza and Aleksy said he ate $\frac{2}{4}$ of his pizza.

Which boys ate the same amount of pizza?

Toby
Ronan
Toby
Ronan [1]

2 Lila made a large cake for her birthday. She cut the cake into 42 equal pieces and 28 guests took a piece home.

What fraction of the cake was taken? Write the answer in its simplest form.

$\frac{28}{42} = \frac{4}{6} = \frac{2}{3}$ [1]

3 To keep herself fit, Iona jogged $\frac{1}{4}$ of a mile every day for a week.

How many miles did she jog in total?

$\frac{1}{4} \times \frac{7}{1} = \frac{7}{4} = 1\frac{3}{4}$ $1\frac{3}{4}$ [1]

4 Jay and Pasha shared a bar of chocolate. Jay ate $\frac{1}{3}$ of the bar and Pasha ate $\frac{2}{3}$.

How much chocolate was left?

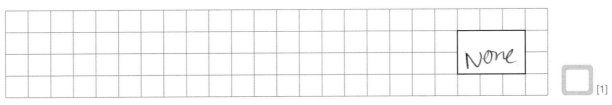

None [1]

01/09
7 min

Decimals

> 💡 **Helpful Hint**
>
> A **decimal fraction** is a **decimal number** that is less than 1. For example: 0.58
>
> A **mixed decimal** is a **decimal number** that is more than 1. For example: 3.58
>
> When adding and subtracting decimals, you must always make sure the **decimal points** are lined up.
>
> **Example:**
>
> | | 6 | 2 | . | 5 | |
> |---|---|---|---|---|---|
> | + | | 3 | . | 2 | 8 |
> | | 6 | 5 | . | 7 | 8 |
> | | | | | | |

Ⓐ Answer these questions. You do not need to show your workings out.

1 0.4 + 0.5 = _0.9_ ☐ [1]

2 1.6 + 0.2 = _1.8_ ☐ [1]

3 0.9 − 0.7 = _0.2_ ☐ [1]

4 1.8 − 0.6 = _1.2_ ☐ [1]

5 2.7 + 2.4 = _4 5.1_ ☐ [1]

6 3.4 − 1.2 = _2.2_ ☐ [1]

7 14.9 + 10.3 = _25.2_ ☐ [1]

8 22.6 − 2.8 = _19.8_ ☐ [1]

9 12.3 + 6.8 = _19.1_ ☐ [1]

10 37.9 − 0.4 = _37.5_ ☐ [1]

☐ / 10

01/09
3 min

Ⓑ **Answer these questions and show your workings out.**

1 23.21 + 36.38 =

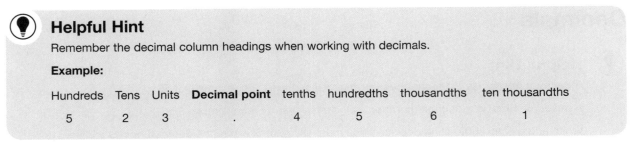

$$\begin{array}{r} 2\,3.2\,1 \\ 3\,6.3\,8\,+ \\ \hline 5\,9.5\,9 \end{array}$$

59.59   [1]

2 255.9 + 21.62 =

$$\begin{array}{r} 2\,5\,5.9\,0\,+ \\ 2\,1.6\,2 \\ \hline 2\,7\,7.5\,2 \end{array}$$

277.52   [1]

3 0.48 + 0.52 =

$$\begin{array}{r} 0.4\,8\,+ \\ 0.5\,2 \\ \hline 1.0\,0 \end{array}$$

1   [1]

4 1026.663 + 28.3496 =

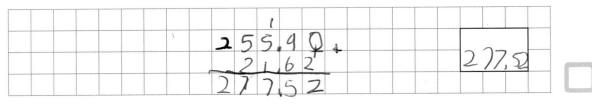

$$\begin{array}{r} 1\,0\,2\,6.6\,6\,3\,0\,+ \\ 2\,8.3\,4\,9\,6 \\ \hline 1\,0\,5\,5.0\,1\,2\,6 \end{array}$$

1055.0126   [1]

5 4569.46 + 7398.29 =

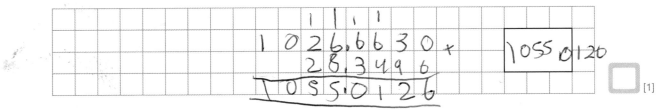

$$\begin{array}{r} 4\,5\,6\,9.4\,6\,+ \\ 7\,3\,9\,8.2\,9 \\ \hline 1\,1\,9\,6\,5.7\,5 \end{array}$$

11965.75   [1]

5

ⓒ Answer these questions and show your workings out.

1 4012.78 – 210.6 =

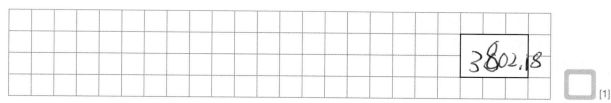

3802.18   [1]

2 24.5 – 9.7 =

14.8   [1]

3 19.52 – 16.37 =

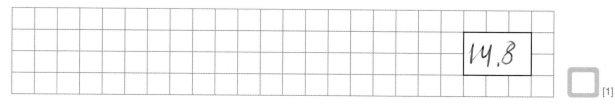

3.15   [1]

4 52 – 48.4 =

3.6   [1]

5 107.3 – 52.9 =

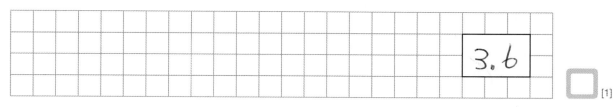

54.4   [1]

6 139 – 0.954 =

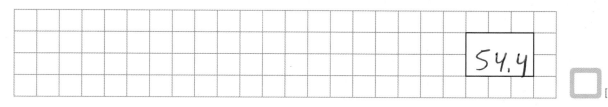

138.046   [1]

6

Word problems

ⓓ Solve these word problems and show your workings out.

1 Becca and Livvy are good swimmers and had a race. Becca beat Livvy by 0.3 of a second. Becca's time was 10.46 seconds.

What time did Livvy get?

[1]

2 Rhys is training for a marathon. As part of his training he did two long training runs. The first run was 14.75 miles, the second was 18.5 miles.

How many miles did he run in total?

[1]

3 23.7 cm was cut from a ribbon measuring 44.5 cm.

How much ribbon was left?

[1]

4 Jess ran 100 m in 15.82 seconds. Her aim is to run it in 14.9 seconds.

How much time does Jess need to save to achieve her target?

[1]

5 Sam buys a book for £14.87.

How much change will get if he pays with a £20 note?

[1]

5

Test your skills

Helpful Hint

The questions in this unit are mixed calculations. Time yourself to see how well you can do. Look back at the other units if you need to refresh your memory.

(A) Answer these questions. You do not need to show your workings out.

1 5.8 + 2.2 = _____ ☐ [1]

2 667 − 87 = _____ ☐ [1]

3 0.3 − 6 = _____ ☐ [1]

4 9^2 = _____ ☐ [1]

5 672 + 328 = _____ ☐ [1]

6 8.6 − 4.7 = _____ ☐ [1]

7 2.7 × 10 = _____ ☐ [1]

8 6093 ÷ 3 = _____ ☐ [1]

9 120 × 7 = _____ ☐ [1]

10 2575 ÷ 100 = _____ ☐ [1]

10

Unit 10

(B) Answer these questions and show your workings out.

1 67 342 − 8348 =

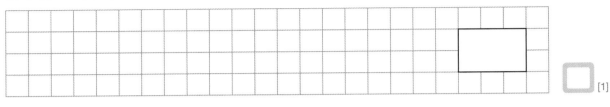

[1]

2 8^3 =

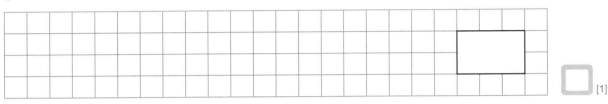

[1]

3 78.34 + 2.88 =

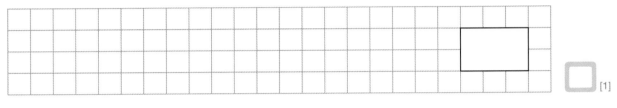

[1]

4 54 899 + 72 377 =

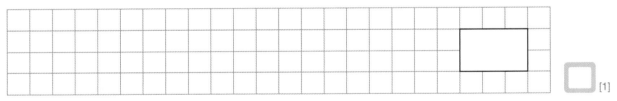

[1]

5 $\frac{1}{8}$ + $\frac{4}{8}$ =

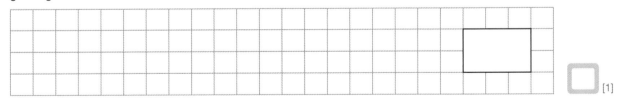

[1]

6 3826.27 ÷ 100 rounded to the nearest 10 =

[1]

6

7 2250 ÷ 7 =

[1]

8 3375 × 16 =

[1]

9 Write $\frac{40}{7}$ as a mixed number.

[1]

10 1407 ÷ 6 =

[1]

11 27.56 + 45.94 =

[1]

12 Write $\frac{36}{54}$ in its simplest form.

[1]

6

Unit 10

Word problems

💡 **Helpful Hint**
Remember there are 1000 m in 1 km.

© Solve these word problems and show your workings out.

1 Liam swam 400 m in 15 minutes. If he continued to swim at the same rate how many kilometres could he swim in one hour?

[1]

2 Miya bought fifteen cans of cat food at 27p each.

How much did the cat food cost in total?

[1]

3 If you divide a number by 3 and the answer is 2361, what is the number?

[1]

4 An ink cartridge prints 450 pages before running out. How many ink cartridges would be needed if 4500 pages were to be printed?

[1]

5 There are 24 children in a class. Three-quarters of them have paid £7.45 each for a school photograph.

How much money has been collected?

[1]

5

Key words

Common factors numbers that fit exactly into two or more larger numbers, for example *7 is a common factor of 7, 35 and 210*

Composite number a number that is not a prime number

Cube number a number that is the result of a number being multiplied by itself then multiplied by itself again, for example *8 (2 × 2 × 2)*

Decimal fraction a fraction that has 10, 100 or 1000 on the bottom that can then be placed on the decimal system using the decimal point, for example $\frac{40}{100} = 0.40$ *(40 hundredths)*

Decimal number a number that uses the decimal point

Decimal point a decimal point comes between the units and the tenths. It separates whole numbers from fractions

Denominator the bottom number of a fraction, which tells you how many equal parts the shape or amount has been divided into

Digit one of the ten symbols: 0, 1, 2, 3, 4, 5, 6, 7, 8, 9

Equivalent fractions a fraction which has the same value as another fraction but has been written with different numbers. When you simplify a fraction, the new fraction is equivalent to the original fraction, for example $\frac{1}{2} = \frac{50}{100}$

Factor a number that fits exactly into a larger number

Factor pairs two numbers that multiply together to make another number

Fraction a fraction is a part of one whole thing

Improper fraction this is a fraction where the numerator is larger than the denominator

Inverse operation we can use inverse operations to check our answer to a calculation, for example *2 + 3 = 5 and 5 − 3 = 2*. The inverse of addition is subtraction. The inverse of multiplication is division

Long multiplication a column method of working out a multiplication sum

Mixed decimal a mixed decimal has whole numbers before the decimal point and numbers after the decimal point

Mixed number a number shown with a whole number and a fraction such as $2\frac{1}{2}$

Multiplier a number that we multiply another number by

Negative numbers negative numbers are numbers less than zero

Number line (positive and negative) this is like a section of a ruler showing a section of numbers

Number sequence a series of numbers that form a pattern, such as 10, 11, 12, 13 or 10, 20, 30, 40

Numerator the top number of a fraction is called the numerator. It tells you how many parts of the total amount have been taken such as **two** thirds or **one** half

Place value tells you the value of each digit in a number, such as units, tenths or hundreds

Prime factor a factor that is a prime number

Prime number a number that only 1 and itself can fit exactly into

Remainder the amount left over when one number does not divide exactly into another

Rounding if the digit to the right of the digit we are rounding to is 0, 1, 2, 3 or 4 we round the number down. If it is 5 or above, we round the number up

Simplest form reducing the numbers in a fraction or ratio until they can no longer be divided by a common factor

Square number the result of a number being multiplied by itself, for example *4 (2 × 2)*

How did you do? Fill in your score below and shade in the corresponding boxes
to compare your progress across the different tests and units.

50%   100%   50%   100%

Unit 1, p3 Score: __ / 9

Unit 1, p4 Score: __ / 7

Unit 1, p5 Score: __ / 6

Unit 1, p6 Score: __ / 6

Unit 2, p7 Score: __ / 10

Unit 2, p8 Score: __ / 5

Unit 2, p9 Score: __ / 6

Unit 2, p10 Score: __ / 4

Unit 3, p11 Score: __ / 10

Unit 3, p12 Score: __ / 4

Unit 3, p13 Score: __ / 6

Unit 3, p14 Score: __ / 4

Unit 4, p15 Score: __ / 10

Unit 4, p16 Score: __ / 4

Unit 4, p17 Score: __ / 6

Unit 4, p18 Score: __ / 5

Unit 5, p19 Score: __ / 12

Unit 5, p20 Score: __ / 4

Unit 5, p21 Score: __ / 4

Unit 5, p22 Score: __ / 5

Unit 6, p27 Score: __ / 10

Unit 6, p28 Score: __ / 5

Unit 6, p29 Score: __ / 5

Unit 6, p30 Score: __ / 6

Unit 7, p31 Score: __ / 10

Unit 7, p32 Score: __ / 6

Unit 7, p33 Score: __ / 6

Unit 7, p34 Score: __ / 5

Unit 8, p35 Score: __ / 12

Unit 8, p36 Score: __ / 5

Unit 8, p37 Score: __ / 4

Unit 8, p38 Score: __ / 4

Unit 9, p39 Score: __ / 10

Unit 9, p40 Score: __ / 5

Unit 9, p41 Score: __ / 6

Unit 9, p42 Score: __ / 5

Unit 10, p43 Score: __ / 10

Unit 10, p44 Score: __ / 6

Unit 10, p45 Score: __ / 6

Unit 10, p46 Score: __ / 5